Child-Centered Math

Geometry

35 Hands-On Activities

Grades 2–3

Written by Cindy Barden
Edited by Janet Bruno
Illustrated by Terri Sopp Rae
Project Director: Carolea Williams

Table of Contents

To the Teacher

The child-centered activities in this book are the perfect way to expand students' knowledge of geometry. These 35 hands-on activities cover the following concepts: plane shapes, sides and angles, solids, symmetry, congruence, perimeter, area, volume, and coordinate geometry. Each activity is quick to prepare and highly motivating, making learning about geometry an enjoyable experience.

Geometry is designed as a handy resource for teachers, not a prescribed continuum of geometry skills. Integrate the activities into your current mathematics program, keeping in mind the special needs of your students.

Have fun watching your students get excited about math as they participate in these interactive geometry activities.

Getting Started

Most activities can be implemented in small- or large-group settings, but some are best suited to learning centers where a few students can work independently. When planning the amount of adult guidance or participation needed, keep in mind the materials to be used, the type of work space needed, and the activity level of the project.

Geometry is ideal for involving teacher aides or parent volunteers in the classroom. The directions are simple and easy to follow, and students will quickly become engaged in the activities. It may be helpful to place a laminated copy of each activity in a box with the materials. This allows for instant setup and cleanup.

Although most activities use a variety of inexpensive, readily-available materials, some parent donations may be helpful. A parent letter is included on page 36 to help you obtain various consumable materials.

About Geometry

Geometry is the branch of mathematics that studies points, lines, angles, surfaces, planes, and solids and how they are related. A knowledge of geometric concepts and relationships is essential in many real-life applications (e.g., figuring quantities of wallpaper or building a bookcase) and occupations (e.g., art, architecture, engineering).

By the time students enter second and third grade, they have had many informal experiences with geometry. They have built a wide variety of structures with blocks (geometric solids), used basic shapes for art projects, and sorted shapes by various attributes. They have discovered shapes and lines in their environment and developed spatial sense doing puzzles.

Second and third graders can expand on these first experiences by using a wide variety of concrete materials to build understanding. In addition to learning to identify plane and solid shapes, students need opportunities to create shapes, compare shapes, and investigate relationships among them. They learn about symmetry and congruence by creating their own symmetrical designs and congruent shapes.

Shapes, Shapes, and More Shapes

Materials

- tagboard or cardboard
- scissors
- chalkboard
- chalk
- paper grocery bag

Procedure

1. Cut various-sized circles, squares, triangles, rectangles, pentagons, hexagons, and octagons from tagboard or cardboard.
2. Working with a small group, have students sort the shapes by type (e.g., all squares, all rectangles, and so on).
3. Hold up two different shapes, and have students name ways they are alike and different.
4. List the characteristics (e.g., number of sides, number of corners or angles, length of sides) and name of each shape on the chalkboard. Tell students that flat figures like triangles, circles, and squares are called *plane* figures. Plane means flat.
5. Place the shapes in a bag. One at a time, have students reach into the bag and try to identify one shape by touch. Ask them to also name the characteristics of the shape.

It has four corners.

Notes:

Extensions:

Shapes Inside of Shapes

Materials

- 6" paper squares
- scissors
- 9" x 12" construction paper
- glue
- markers

Procedure

1. Give each student several squares.
2. Tell students to experiment with folding the squares so one has four equal squares, one has four equal rectangles, and one has four equal triangles.
3. Ask students to cut apart each square on the folds and use the shapes to make shape cartoon characters.
4. Encourage them to use markers to add details and speech bubbles for each character.

Extension Begin with paper rectangles, triangles, octagons, and hexagons. Have students fold the shapes in different ways to create new shapes.

Notes:

Extensions:

Mix Them Up

Materials

- construction paper circles, squares, triangles, rectangles, hexagons, octagons, and pentagons
- 12" x 18" construction paper
- glue
- markers

Procedure

1. Demonstrate how shapes can be combined to make designs and pictures.
2. Give each student glue, construction paper, markers, and shapes of various sizes.
3. Ask students to make designs or pictures using the shapes. Designs could represent real objects such as a house or an ice cream cone, or fanciful "objects" such as a sneeze or a hiccup.
4. Encourage students to manipulate the shapes into a variety of designs before deciding on a final plan. Then have them glue the shapes to construction paper and add details with markers.
5. Ask students to title their art projects. Display student art on a bulletin board.

Extension Provide students with sets of tangrams. Have them make different figures using all seven shapes.

Notes:

Extensions:

Shape Hunt

Materials

- paper
- pencils

Procedure

1. Divide the class into partners. Give each student pair paper and pencils.

2. Have students divide their papers into six sections and label the sections *Squares, Rectangles, Circles, Triangles, Ovals,* and *Octagons.*

3. Tell students to search for shapes in various areas of the school and school grounds. Encourage them to look at walls, ceilings, floors, furniture, clothing, and parts of objects that may contain several different shapes.

Squares	Rectangles	Circles	Triangles	Octagons	Ovals
window 4-square court	door table book	clock tire		stop sign	table

4. When they spot a shape, have students write the object's name in the appropriate section. For example, they would write *back of chair* under *Rectangles* and *bottom of cup* under *Circles.*

5. When the Shape Hunt is over, have students tally the number of shapes they found. Which shape was the most/least common?

Notes:

Extensions:

Puzzle Shapes

Materials

- manila envelopes
- poster board or tagboard
- crayons
- scissors
- rulers
- pencils

Procedure

1. Cut various shapes from poster board or tagboard.
2. Give each student an envelope and one shape. Have each student write the name of the shape on the envelope and color one side of the shape.
3. With a pencil and ruler, have students draw straight lines on their shapes to divide them into five or six sections.
4. Have students cut along the lines to make a puzzle, then place the pieces in the envelope.
5. Ask students to exchange envelopes with a classmate and solve the other student's puzzle by arranging the pieces into the original shape.

Extension Using the same puzzle pieces, ask students to rearrange them to make other multi-sided figures.

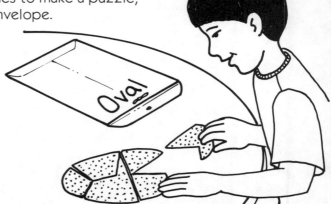

Notes:

Extensions:

Geometric Patterns

Materials

- items with geometric shapes and repeating patterns (fabric, checkerboard, wallpaper samples, quilt pictures)
- graph paper
- colored pencils or crayons

Procedure

1. Show students samples of objects with geometric patterns.
2. Have students describe the patterns in terms of shape and color.
3. Ask students to create their own geometric patterns on graph paper using colored pencils. Remind them that any combination of geometric shapes and any colors can be used as long as they form a pattern.

Variation Have students make patterns by gluing shape cutouts on construction paper.

Notes:

Extensions:

Tessellations

Materials

- artwork samples by M.C. Escher
- 3" tagboard squares
- scissors
- tape
- 12" x 18" construction paper
- markers

Procedure

1. Show the class samples of M.C. Escher's tessellations. Tell them they will create similar tessellations with interlocking shapes.
2. Give each student several tagboard squares.
3. Have students cut a shape from one side of the square and tape it to the opposite side. Let them experiment with several squares.
4. Show students how to trace around their shapes on construction paper, making sure the pieces interlock. There should be no gaps or blank spaces.
5. When the whole sheet is covered, invite students to decorate their tessellating designs.

Notes:

Extensions:

One More Side

Materials

- dot paper
- pencils
- rulers

Procedure

1. Have students draw a three-sided polygon on dot paper.
2. Challenge them to continue in a pattern, drawing a four-sided figure, a five-sided figure, and so on.
3. Ask students to share their work and discuss the variety of polygons created.

Extension Teach students the mathematical name for each shape, such as triangle, quadrilateral, pentagon, hexagon, and so on.

Notes:

Extensions:

Octominoes

Materials

- Color Tiles or 1" paper squares
- pencils
- paper

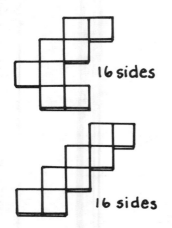

16 sides

16 sides

Procedure

1. Divide students into partners. Give each pair about 40 Color Tiles.

2. Have students arrange eight tiles in any manner as long as one whole side of each square touches one whole side of another square. This is an "octomino."

This one has 16 sides.

3. Challenge students to build an octomino with the most possible sides. (Several different arrangements result in a 16-sided figure.) Give them time to make several octominoes.

4. Have students trace around each octomino on paper. Have them count the number of sides and write that number in the figure.

5. Invite students to compare their octominoes with those made by other pairs. Point out how figures with the same number of sides can have different shapes.

Notes:

Extensions:

Right Angles All Around

Materials

- objects with right angles (book, box, paper)
- 5" x 5" tagboard squares
- plane figures (rectangles, triangles, parallelograms, trapezoids, hexagons)

Procedure

1. Use objects to show students several examples of right angles. Since all square corners are right angles, lay a tagboard square against the right angle of each object.

2. Ask students to find examples of right angles in their classroom (e.g., where a wall meets the floor). Students can check for right angles by holding the corner of a tagboard square against another angle to see if they match.

3. Ask students to find examples of angles that are larger than (obtuse angle) and smaller than (acute angle) a right angle. Have them test and compare using tagboard squares.

4. Provide students with plane figure models. Ask them to compare the right angles on their tagboard squares with the angles in each shape. Which are right angles? Which are larger/smaller?

It's a right angle.

Math

Notes:

Extensions:

Around a Circle

Materials

- chalk
- chalkboard
- ruler
- compasses
- drawing paper
- markers

Procedure

1. Draw a circle on the board. Put a dot in the center. Explain that to be a circle, every point along the outside of the circle must be exactly the same distance from its center.

2. Ask a student to measure the distance from the center of the circle to the edge in several places with a ruler. Is the distance the same?

3. Demonstrate how to draw a circle with a compass. Let students practice drawing circles of different sizes.

4. Give students another piece of paper and have them use the compass to create circle designs. Encourage them to experiment and overlap circles.

5. Have them enhance designs by filling in different sections with patterns or solid blocks of color.

Extension Introduce the words *radius, diameter,* and *circumference* with this activity.

Notes:

Extensions:

Circle Time

Materials

- chalk
- various lengths of string

Procedure

1. Use a large paved area for this activity, such as the playground or a sidewalk.
2. Have students work in pairs. Give each pair a piece of chalk and string.
3. Have students tie the string to the chalk and make a dot for the center of the circle.
4. Have one student hold the string in place at the dot while the other draws a circle with the chalk. Remind students to keep the string taut as they draw.
5. When they finish, discuss why the circles are different sizes. What happens when you use a longer piece of string? a shorter piece?

Notes:

Extensions:

Round and Round

Materials

- paper circle
- spherical objects (tennis ball, marble, globe, basketball, ball of yarn)
- assorted circular and cylindrical objects and pictures (penny, yo-yo, paper plate, cake, pencil)
- modeling clay

Procedure

1. Hold up a paper circle and a ball. Ask students to name ways the two objects are alike and different. Tell students that three-dimensional objects like baseballs are called *spheres*.
2. Place spherical and non-spherical objects in a box. Hold up one object at a time and ask the class if it is a sphere. Ask them to explain their answers.
3. Give each student a piece of modeling clay. Have them use the clay to make a sphere, a flat circle, and a cylinder.
4. Ask students to make a list of at least ten spherical objects.

Extension Use a globe to demonstrate the word *hemisphere*. "Hemi" means half.

Notes:

Extensions:

Butterfly Symmetry

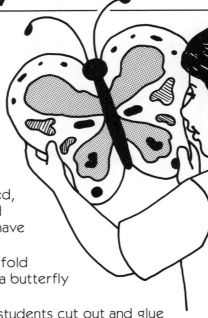

Materials

- tempera paints
- paintbrushes
- colored butcher paper
- black construction paper
- scissors
- pencils
- glue

Procedure

1. Place materials at a center. Show students how to make butterflies with symmetrical wing patterns.
2. Have students fold a piece of butcher paper in half and dab blobs of paint on one side only.
3. Tell them to fold the paper closed, gently press on the top side, and open the paper again. They will have a symmetrical pattern.
4. When the paint is dry, have them fold the paper and draw and cut out a butterfly wing shape on the fold.
5. To complete the butterfly, have students cut out and glue on a body and two antennae from black construction paper.

Extension Have students create symmetrical fantasy insects using the same process.

Notes:

Extensions:

Symmetry Puzzles

Materials

- construction paper
- scissors
- paper grocery bag

Procedure

1. Ask students to look for symmetrical objects in the classroom. Objects like an open book, a pair of glasses, and a sweater are symmetrical because both sides of the object are exactly the same.

2. Have students cut symmetrical figures by folding a sheet of paper in half and cutting out a shape on the fold. When the paper is unfolded, the shape will be symmetrical—both halves will be exactly the same.

3. Have students cut their shapes in half along the fold, place pieces in a bag, and mix them up.

4. Ask each student to take two pieces from the bag. Have them find classmates with matching symmetrical halves.

Extension Invite students to cut paper snowflakes as another way to make symmetrical figures. Have them trace all the lines of symmetry.

Notes:

Extensions:

Lines of Symmetry

Materials

- 10–12 construction paper shapes
- chalkboard
- chalk

Procedure

1. Cut the shapes in half from various angles so some halves are symmetrical and some are not.
2. Working with a small group, demonstrate a line of symmetry by drawing a heart on the chalkboard. Draw a vertical line through it so both halves are exactly the same.
3. Draw other symmetrical shapes on the board (equilateral triangle, pentagon, circle, three-leaf clover). Have students take turns drawing a line (or lines) of symmetry on the shapes.
4. Distribute the shapes. Have students match the two halves.
5. Have students sort the shapes into two piles—those cut on a line of symmetry and those that aren't.

Notes:

Extensions:

Symmetrical Letters

Materials

- writing paper
- pencils
- small mirrors

Procedure

1. Ask students to print the alphabet neatly in capital letters. Tell them to leave about an inch between each letter.
2. Have students circle all letters that are symmetrical. (A, B, C, D, E, H, I, K, M, O, T, U, V, W, X and Y are symmetrical.)
3. Ask students to draw a line of symmetry through each symmetrical letter. A line of symmetry divides a letter or an object in half so both halves are exactly the same.
4. Have students test each line of symmetry by placing a small mirror on the line. If the line of symmetry is correct, the letter will look whole.

Variation Do the same activity with numerals *1–11*.

Extension Many items in nature are symmetrical. Have students gather symmetrical leaves, seed pods, and other objects. Ask them to trace around the objects on paper and draw lines of symmetry.

Notes:

Extensions:

Geoboard Game

Materials

- 4–6 geoboards
- rubber bands
- paper
- pencils

Procedure

1. Place materials at a center, and have students work in pairs.
2. Tell Player A to watch as Player B uses rubber bands to build a polygon on a geoboard.
3. Have Player B cover the shape, and ask Player A to make a congruent shape from memory on a separate geoboard. If successful, players switch roles and repeat the activity.
4. Each time a player builds a congruent shape, he or she gets one point.
5. The player with the most points at the end of a designated time wins.

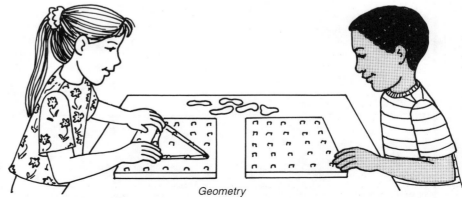

Notes:

Extensions:

Congruent Shapes

Materials

- Color Tiles or 1" paper squares
- file folders

Procedure

1. Divide the class into partners. Give each student pair some Color Tiles and a file folder.
2. Have students stand up the file folder between them.
3. Ask one student to build a simple shape with Color Tiles on one side of the file folder.
4. Ask the same student to give verbal directions to his or her partner on how to build a congruent shape (e.g., "Place three tiles in a horizontal row. Add one tile above the middle tile . . . ").
5. When his or her partner thinks the shape is correct, tell students to remove the divider and check.
6. Have students switch roles and repeat the activity several times.

Place a yellow tile directly above the center blue tile.

Notes:

Extensions:

Match It

Materials

- dot paper
- pencils
- rulers

Procedure

1. Have students work in pairs. Ask one student to draw a polygon on dot paper, using a ruler to make certain lines are straight.
2. Have his or her partner draw the same figure, exactly the same size and shape, on a separate piece of dot paper. Tell students to count dots for accuracy.
3. Have students compare their shapes. If the two shapes are exactly the same, the shapes are congruent.
4. Ask students to repeat steps 1–3 several times, taking turns drawing first.

Extension Have students make congruent figures by folding paper and cutting paper dolls, trees, or other shapes.

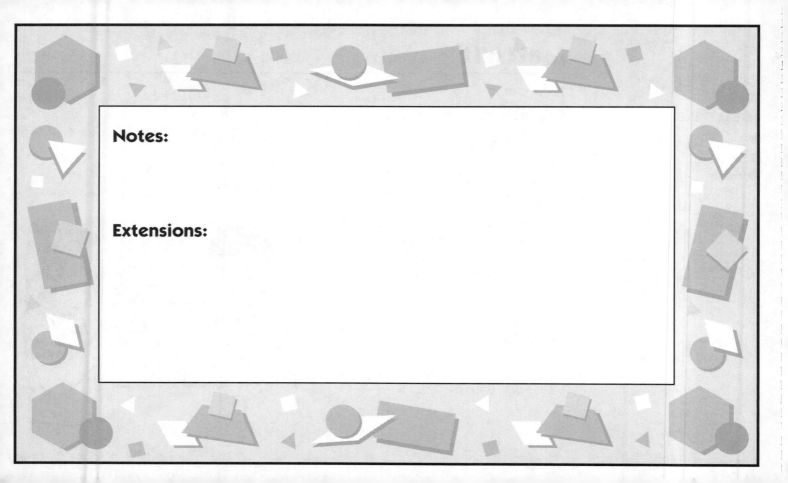

Notes:

Extensions:

Congruent Memory Game

Materials

- unlined 3" x 5" index cards
- construction paper scraps
- scissors
- glue

Procedure

1. Work in groups of four to five students. Give each group about 20 cards.
2. Have students cut congruent shapes from construction paper and glue them on the cards. Show students how to fold the paper and cut two shapes at one time to make sure the shapes are congruent.

3. Have students gather the cards, shuffle them, and spread them out face down in rows.
4. Have one student turn two cards face up. If the figures are congruent, tell the student to keep the pair and turn over two more cards. If the figures are not congruent, tell the student to turn the cards face down, and have the next student take a turn.
5. Play continues until all pairs are matched. The player with the most cards wins.

Notes:

Extensions:

Solids Scavenger Hunt

Materials

- solid shapes: cubes (dice, wooden blocks), cylinders (paper towel tubes, straws), pyramids (wooden models), cones (ice cream cones), spheres (balls, marbles), and rectangular solids (small boxes)
- 6 boxes or bags
- blindfold

Procedure

1. Label the boxes *Cubes, Cylinders, Pyramids, Cones, Spheres,* and *Rectangular Solids.*
2. Explain that *solid* figures are different than *plane* figures. Solid figures are three-dimensional, and plane figures are two-dimensional.
3. Show students the solid figures, and have them sort the figures by type into the boxes.
4. Ask students to look around the room to find other geometric solids to include in the collection.
5. Have students continue the Solids Scavenger Hunt at home. Ask them to bring objects to school to share with the class.
6. Place materials at a center. Blindfold one child and have students take turns guessing shapes by touch.

It's a cylinder!

Notes:

Extensions:

3-D Sculptures

Materials

- miniature marshmallows
- round toothpicks

Procedure

1. Divide students into small groups. Give each group marshmallows and toothpicks.

2. Challenge students to make various plane and solid shapes, using toothpicks as sides and marshmallows as corners.

3. After experimenting for a time, have students make specific shapes. Start with plane shapes such as triangles and squares, then progress to solid shapes such as cubes, rectangular solids, and pyramids.

Extension Have students build houses and other buildings.

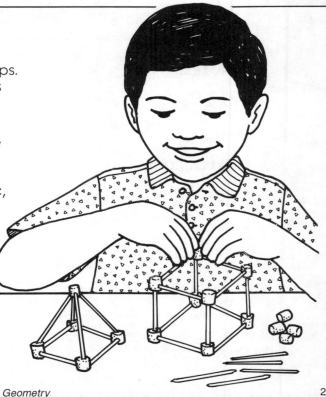

Notes:

Extensions:

Pyramid Power

Materials

- Pyramid Pattern reproducibles (pages 37 and 38)
- construction paper or tagboard
- scissors
- tape
- crayons or markers

Procedure

1. Enlarge the pyramid patterns and reproduce one for each student on construction paper or tagboard.
2. Ask students to look at the patterns and predict what 3-D figures they will make.
3. Have students cut on the solid lines and fold on the dotted lines. Invite students to decorate their shapes before final assembly.
4. Have students work in pairs to tape together the pyramid sides.
5. For each figure, have students count the number of corners, edges, square faces and/or triangular faces.

Extension Show students pictures of the Egyptian pyramids. Have students research about their construction.

Notes:

Extensions:

Building Solids

Materials

- cube, cylinder, and cone models
- construction paper
- graph paper
- tape
- scissors
- pencils

Procedure

1. Ask students to study the cylinder, then work with a partner to make a cylinder using construction paper and tape. Challenge them to think of a way to close off each end of the cylinder.

2. Ask students to study the cone, then make one using construction paper and tape. Challenge them to close the base of the cone.

3. Ask students to study the cube and imagine how it would look if the sides were unfolded into one flat piece.

4. Give each pair of students several sheets of graph paper. Have them experiment making a pattern for a cube. Have students draw a pattern on graph paper, and then cut, fold, and tape the sides together.

Notes:

Extensions:

What Am I?

Materials

- index cards
- markers
- manila envelope
- straight pins

Procedure

1. Write names of geometric shapes (e.g., *square, triangle, octagon, sphere, cube*) on index cards. Make one card for each student.

2. Put the cards in a large envelope.

3. Pin one card on the back of each student. Don't let students see the words on their cards.

4. Invite students to ask classmates questions to determine the names of the shapes. Have them ask only "yes" or "no" questions (e.g., "Am I a plane figure?" "Do I have less than four sides?" "Am I round?").

5. As students guess their shapes, have them return the cards to a central location and continue answering other students' questions.

Notes:

Extensions:

Geometric Poetry

Materials

- construction paper
- scissors
- pencils
- crayons, colored pencils, markers

Procedure

1. Have each student cut construction paper into a geometric shape of his or her choice.

2. On the shapes, ask students to write a cinquain (a five-line poem) using this formula:

   First line: Name of the shape.

   Second line: Two words that describe the shape.

   Third line: Three words naming objects of that shape.

   Fourth line: Two more words describing the shape.

   Fifth line: Name of the shape.

3. Have students decorate their poems and display them on a bulletin board.

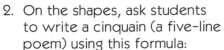

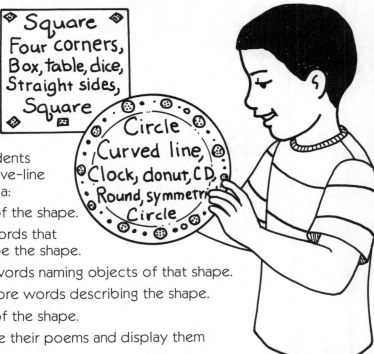

Square
Four corners,
Box, table, dice,
Straight sides,
Square

Circle
Curved line,
Clock, donut, CD,
Round, symmetri
Circle

Notes:

Extensions:

Around the Outside

Materials

- book
- 3' pieces of string
- yardsticks or meter sticks
- paper
- pencils

Procedure

1. Show students how to measure the perimeter of a book with a piece of string and how to get a more standard measurement by matching the string to a yardstick or meter stick.
2. Divide the class into partners. Give each student pair a 3' piece of string and a yardstick or meter stick.
3. Have students measure the perimeter of geometric solids in the classroom (e.g., top of desk, eraser, small chalkboard). Ask them to match the string to the yardstick or meter stick and record their measurements.
4. Challenge students to find objects with a particular perimeter such as 12 inches or 60 centimeters.

Extension Ask students to measure the perimeter of irregular-shaped objects.

Notes:

Extensions:

Count Your Way Around

Materials

- nonstandard measuring tools (straws, sticks, blocks, books, shoes)
- chalk
- large paved surface
- paper
- pencils

Procedure

1. Divide the class into small groups. Give each group member a different nonstandard measuring tool.
2. Go outside and use chalk to draw a large geometric shape for each group.
3. Have each group member measure the perimeter of the shape using a different tool.
4. Ask one child in each group to record the results. Encourage students to discuss why their perimeter measurements were different. Ask students to choose the most efficient measuring tool.

Extension Have students measure the perimeter of the classroom. Let them choose an appropriate measuring tool.

Notes:

Extensions:

Measuring Area

Materials

- paper or cardboard squares (12", 6", 3", 1")
- paper
- pencils
- graph paper

Procedure

1. Have students work together to measure the area of various objects in the classroom (door, window, desktop) by covering them with paper squares. Encourage them to choose the appropriate square size for the object being measured.
2. Ask students to record each object and write the number of square units it took to cover the object.
3. Using their notes, have students draw the objects proportionately on graph paper.

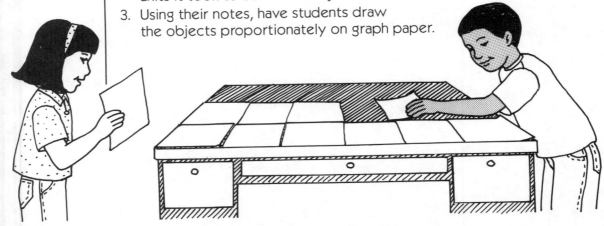

Geometry

Notes:

Extensions:

Same Area, Different Shapes

Materials

- graph paper
- pencils
- crayons, colored pencils

Procedure

1. Tell students to color in any eight adjoining squares on the graph paper.
2. Have students compare the shape they colored to those colored by classmates. Ask students what the figures have in common. (All cover the same area.)
3. Have students make three other shapes with an area of eight and demonstrate how different shapes can cover the same area.

Extension Ask students to name products sold according to the area they cover (e.g., carpeting, grass seed, paint, tiles).

Notes:

Extensions:

Volume

Materials

- linking cubes or small blocks
- small boxes of varying sizes (shoe boxes, empty cereal boxes)

Procedure

1. Work in small groups, and give each student a box.
2. Ask students to estimate how many cubes will fit into their boxes.
3. Have students neatly stack the cubes in their boxes, then count the number of cubes used.
4. Have students compare their estimates with the actual number.
5. Tell students to use the same number of cubes and stack them in a different shape outside the box. Ask students to find a connection between the cubes stacked in the box and the second stack. (The volume of the two are equal.)

Notes:

Extensions:

Estimating Volume

Materials

- plastic containers of different sizes and shapes
- pitcher of water
- tray or plastic dishpan

Procedure

1. Display various containers to a small group of students.
2. Hold up two containers and ask students to estimate which container will hold more water.
3. Have a student fill one container with water, and then pour it into the second container. Was their estimate correct?
4. Continue with a different pair of containers until all the containers have been used at least once.
5. Ask students to explain what they learned about how shape and size relate to volume.

Variation   Do steps 1–4 of this activity in a sandbox, using sand instead of water.

Notes:

Extensions:

Coordinate Tic-Tac-Toe

Materials

- overhead transparency of Coordinate Grid
- overhead projector
- overhead markers
- Coordinate Grid reproducible (page 39)
- crayons or markers

Procedure

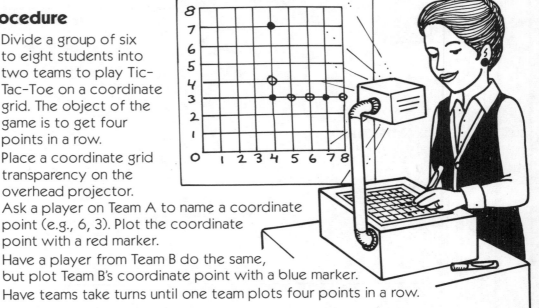

1. Divide a group of six to eight students into two teams to play Tic-Tac-Toe on a coordinate grid. The object of the game is to get four points in a row.

2. Place a coordinate grid transparency on the overhead projector. Ask a player on Team A to name a coordinate point (e.g., 6, 3). Plot the coordinate point with a red marker.

3. Have a player from Team B do the same, but plot Team B's coordinate point with a blue marker.

4. Have teams take turns until one team plots four points in a row.

Extension Reproduce the Coordinate Grid for students, and have them play the game in groups of three (two players and a recorder).

Notes:

Extensions:

Coordinate Geometry

Materials

- Coordinate Grid reproducible (page 39)
- pencils

Procedure

1. Divide the class into partners, and give each student a Coordinate Grid. Tell students they will use the grids to play a game about a lost puppy.

2. Have Partner A place a dot on the grid "where the puppy can be found," but keep the location a secret.

3. Tell Partner B to locate the puppy by guessing coordinates and charting them on a grid. For example, "I've lost my puppy. Have you seen her at (5,6)?"

4. Have Partner A give clues to assist in the search. (e.g., "Your puppy isn't here, but I saw her run down, and then go to the left.") Have Partner B use the clues to make more educated guesses until finding the puppy.

5. When the puppy is "found," have partners switch roles.

Notes:

Extensions:

Parent Letter

Dear Parents,

In our child-centered mathematics program, we are exploring geometry. A variety of materials will be used to make the activities meaningful and fun. Some of the items are consumable. If you would be willing to send in objects you have around the house or to purchase an inexpensive item for the classroom, please sign and return the bottom portion of this letter. I will then return the form to you stating the item and the date it is needed. Thank you!

Sincerely, _____

_____

Yes, I would like to contribute an item for classroom geometry activities.

_____
        (signature)

Please send in _____ by _____.

Thank you very much!

Pyramid Pattern A

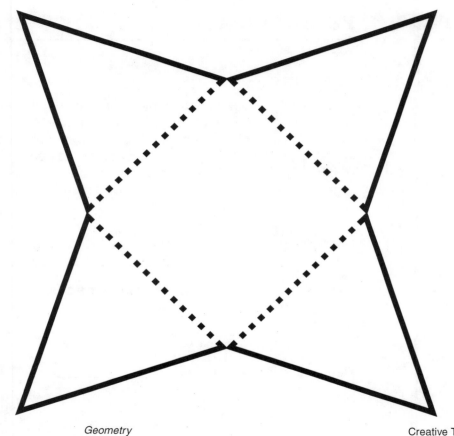

Geometry

Creative Teaching Press, Inc.

Pyramid Pattern B

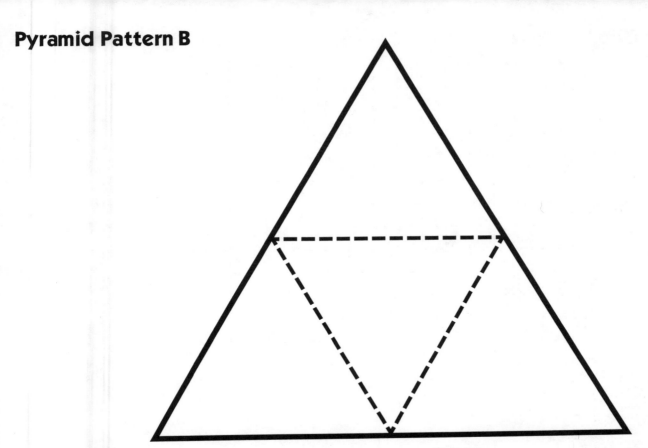

Geometry

Coordinate Grid

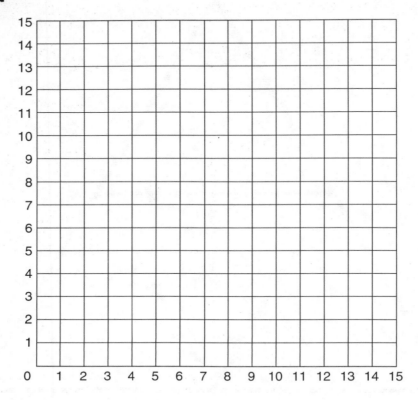

Geometry